Who Am I?

Baby Animals

Written by Read With You
Center for Language Research and Development

The characters and events portrayed in this book are fictitious. Any similarity to real persons, living or dead, is coincidental and not intended by the author.

Published by Read With You Publishing

Designed by Read With You Center for Language Research and Development

ISBN-13: 978-1-944710-87-3
ISBN-10: 1-944710-87-6

Third Edition July 2019
Printed in the United States of America.

I’m born with long legs.

Who am I?

foal

My eggshell has tiny pores that allow me to breathe.

Who am I?

duckling

I'm born blind.

Who am I?

bear cub

I can walk just minutes after birth.

Who am I?

lamb

I break the eggshell with my egg tooth, a sharp bump on my beak.

Who am I?

chick

All of us are born with blue eyes.

Who am I?

kitten

My mom sings to me while nursing.

Who am I?

piglet

I lose my first set of teeth.

Who am I?

puppy

Foal

Duckling

Bear Cub

Lamb

I am a foal.

I am a duckling.

I am a bear cub.

I am a lamb.

Chick

Kitten

Piglet

Puppy

I am a chick.

I am a kitten.

I am a piglet.

I am a puppy.

We are baby animals.

Some animals lay eggs.
Some animals give birth to
their babies.

Made in the USA
Monee, IL
05 November 2026

67931636R00017